Acknowledgements

Research contractor Ove Arup and Partners

Authors

Tim Chapman, BE MSc DIC CEng MIEI FICE

Tim Chapman is a director of Arup and leader of its specialist geotechnical and tunnelling group in London. Tim's experience encompasses major underground building and infrastructure projects all over the world. He has a strong interest in soil-structure interaction problems and the reduction of geotechnical risk. His interest in finding practical and economical solutions to complex foundation problems led to his proposals for foundation reuse, as set out in this guide and in the RuFUS handbook.

Sara Anderson, BA MEng CEng MICE

Sara Anderson is an associate with Arup and has 10 years of experience in the design and construction of underground structures and foundations for building and infrastructure projects. Her experiences, particularly within the congested confines of London, have developed her interest in foundation reuse and the importance of good guidance to overcome current barriers and misconceptions.

Jan Windle, BEng MSc DIC MICE CEng FRSA

Jan Windle, formerly with Arup, is an associate with whitbybird and has around 15 years experience in geotechnical design and construction of both building and civil engineering projects. Jan's key experiences include the analysis and design of sub-structures, foundation systems and ground treatment and she currently acts as an industry reviewer on several projects in promoting the reuse of foundations.

Steering group chairman Hugh St John   GCG

Steering group

Rachel Monteith	Sir Robert McAlpine
Tony O'Brien	Mott MacDonald
Hilary Skinner	BRE
Martin Pedley	Skanska

CIRIA manager CIRIA's research manager for this project was Dr A J Pitchford.

Technical contributors CIRIA and the authors gratefully acknowledge the technical advice and help given by the following individuals:

Jenny Baster	Arup
Juliet Bird	Arup
Bryan Marsh	Arup
Melanie Rhodes	Arup
Ian Rogers	Arup
Jon Shillibeer	Arup
Jim Williams	English Heritage

They would also like to thank Museum of London Archaeology Service for the supply of photographs <www.molas.org.uk>.

Contents

CIRIA C653   London, 2007

Reuse of foundations

T Chapman   Arup

S Anderson   Arup

J Windle   whitbybird (formerly of Arup)

ng best practice

don EC1V 9BP

TELEPHONE 020 7549 3300   FAX 020 7253 0523
EMAIL enquiries@ciria.org
WEBSITE www.ciria.org

Summary

This guide is wholly funded by CIRIA Core programme sponsors and was prepared by a team led by Tim Chapman at Arup under CIRIA Contract 118. The book provides concise guidance for the economic, safe and appropriate reuse of foundations in urban environments where existing foundations are often encountered during redevelopment of buildings and infrastructure.

Reuse of foundations

Chapman, T; Anderson, S; Windle, J

CIRIA

CIRIA C653   © CIRIA 2007   CON118

ISBN-13: 978-086017-653-4
ISBN-10: 0-86017-653-3

British Library Cataloguing in Publication Data

A catalogue record for this book is available from the British Library.

Keywords		
Ground engineering, construction process, sustainability, materials, health and safety		

Reader interest	Classification	
Owners, insurers, developers, designers, geotechnical engineers, structural engineers	AVAILABILITY	Restricted
	CONTENT	Guidance document
	STATUS	Commitee-guided
	USER	Construction professionals (and their advisors) involved with foundation reuse

Published by CIRIA, Classic House, 174–180 Old Street, London EC1V 9BP, UK.

List of figures

List of tables

List of case studies

Executive summary

The reuse of foundations for a second superstructure is technically feasible and is increasingly becoming part of standard practice. For refurbishment projects, reuse of old foundations and structures is the norm.

For foundation reuse to be viable, the following conditions need to apply:

- there should be compatibility between the locations of the applied loads and the existing foundations which should have sufficient capacity to carry the new loads

- sufficient verification should be carried out so that the old foundations are shown to be as reliable as new ones

- there should be an expectation that the foundation performance over the range of expected loads will be acceptable, and that they will fulfil those functions reliably over the planned design life of the building

- the project team need to agree that all parties accept the risks associated with foundation reuse

- adequate insurance cover is available for the design team and client

- regulatory approval is possible from the necessary authorities.

Currently old foundations tend only be reused in a redevelopment if there is a particular constraint that acts as a driver:

- the ground beneath the building has already been filled

- there are archaeological remains that can be preserved by foundation reuse.

One of the main inhibitors to foundation reuse is uncertainty: unless records have been kept that indicate the foundation locations, sizes and capacities with a high degree of reliability, it can be difficult to reuse them reliably or efficiently. Therefore the key to maximising the future ability to reuse foundations is the collection and safe preservation of construction and maintenance records.

Foreword

This guide sets out the background and provides guidance on the key issues for foundation reuse. It is aimed at developers, funders and technical professionals who are interested in an overview of the principal issues relating to reuse of existing foundations for new structures.

The content addresses the practical and technical problems of reusing foundations and suggests strategies for reducing risks.

1   Introduction

1.1   Key issues

Purpose of this CIRIA guide

The purpose of this CIRIA guide is to set out the background and key issues for foundation reuse, in response to an identified need for better dissemination of current knowledge. It is intended to be a concise scoping manual for those considering foundation reuse and not a detailed instruction manual.

> The emphasis is on **pile** reuse, as that has been identified as a common problem with potentially complex technical and commercial issues to resolve. Reuse of shallow foundations is less frequent, but many of the principles are the same.

The guide reviews the practical and technical problems of reusing old foundations and suggests strategies for reducing risks. It is aimed at developers, funders and technical professionals who are interested in an overview of the principal issues relating to reuse of existing foundations for a new building. Its main aim is to demystify the reuse of old foundations while encouraging responsible consideration of the factors that differ from the design and installation of new foundations. A subsidiary aim is to facilitate easier reuse of foundations in the future by encouraging better collection and preservation of records from current piling projects.

The layout of this guide follows the decision-making processes involved in reusing foundations as shown in Figure 1.1.

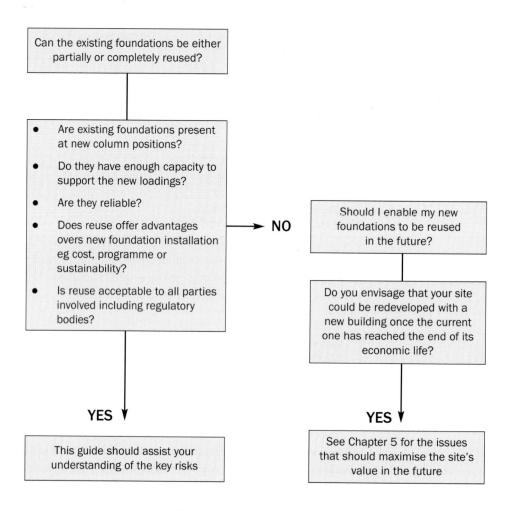

Figure 1.1   *Guide layout*

It has generally been assumed throughout this guide that the redevelopment is led by a commercial developer, who will act as head client and take client decisions. There will be many times when the role of head client is instead taken by a building occupier or another body, who may have different priorities (see Section 3.8). Throughout this guide, "developer" has been used as the generic term for the organisation leading the redevelopment and taking head client decisions.

Does this guide cover all types of foundation?

This guide addresses the potential for reusing all types of foundations. However, shallow foundations are often more easily removed so there is seldom the same difficulty in installing new foundations. Sustainability and waste minimisation potential will encourage consideration of the reuse of all foundation types. Most of the research effort however has been directed at the particular problems of reusing piles, as options are fewer on sites congested with old deep foundations.

Why consider foundation reuse?

Typically for major buildings, a new set of foundations is installed for each new construction. In city centres where property value is high and structures are replaced relatively often, available space for new building foundations is rapidly diminishing. This problem will worsen with time.

In addition to current and future ground congestion, pile reuse is increasingly considered due to:

- potential cost and programme savings by using what would otherwise just be another obstruction
- an increased awareness of the need to protect archaeology undisturbed by the previous foundation schemes
- trends towards greater consideration of sustainability and waste minimisation.

What are other options to foundation reuse?

New foundations can be installed which avoid old foundations. Where it is possible to reuse some of the existing foundations, a mixed foundation solution may be adopted with new foundations used to supplement the reuse of existing foundations. Existing foundations can be removed to reduce the extent of obstructions to new foundations, but this is expensive (see Section 1.3).

What are the cost and programme implications of reusing foundations?

Cost and programme implications will depend on the type of project, the particular site constraints and the viability of using other foundation options. On projects where reuse of old foundations is an option, the cost and programme implications should be assessed at inception and throughout the project. Where good information is available on the existing foundations and the layout is compatible with the proposed redevelopment, cost and programme savings can be made by adopting foundation reuse (see Sections 1.4 and Chapter 3).

Will existing foundations provide a solution as good as installing new?

Foundation systems, whether new or reused, need to function reliably. If existing foundations are to be reused, then the investigation and design process needs to be extended to provide a solution that is as robust as installing new foundations. Reusing foundations in a way that may be less reliable should not be considered as an option (see Chapter 2).

Managing the issues of foundation reuse

Like all construction processes, there are issues that need to be resolved when reusing foundations. It is important that geotechnical and structural engineers work together from the start of a project to set up a comprehensive decision-making strategy to manage the issues so that these risks are no greater than for installing new foundations.

Is foundation reuse a new idea?

No, foundation reuse has occurred in various forms throughout history (see Section 1.2).

What can I do to reduce this problem in the future and maximise the resale value of my site?

The best way to avoid the stagnation of urban sites in the future is to install foundations that can easily be used for successive structures. This means that comprehensive as-built records should be kept including records that explain non-conformances and their assessed significance. Once compiled, these records should be stored in a secure location by the building owner (such as with the CDM Health and Safety file – see Section 5.1) so that they will not be lost due to future events such as the designer or contractor going out of business.

In some cases it may be possible to install a foundation that will support multiple superstructures (see Section 5.3).

What about basements?

Basement retaining walls are even more difficult and costly to remove than individual piles. The ground they retain outside the building is often sensitive to movement, due to adjacent foundations, buried services around the perimeter of buildings and the proximity to tunnels.

Normally the most practical way of retaining the ground outside the basement is to maintain the old retaining wall, at least as part of the temporary works for the new basement. The old wall can then be incorporated into the new permanent works, or be abandoned. If a new wall is installed, for example to allow the new basement to be dug deeper, it is typically constructed inside the line of the old wall as shown in Figure 1.2. Progressively, generations of new walls brought inside the previous ones will lead to a gradual reduction in available basement area. This is an issue for urban underground planning and is outside the scope of this particular guidance.

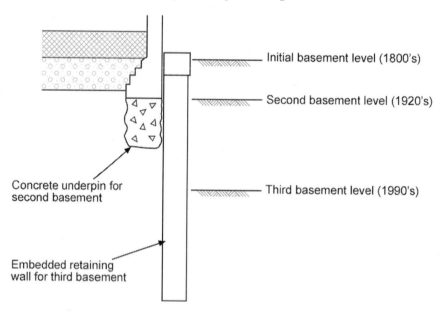

| Figure 1.2 | *Basement depths increasing with time and improvements in construction technology* |

Basement walls are frequently used to carry vertical loads from buildings. Although not explicitly discussed further in this book, the principles of foundation reuse would apply to basement walls acting as foundations.

Where do I go for more technical advice?

Technical advances in foundation reuse were the aim of a European Union funded project on *Reuse of foundations on urban sites* (RuFUS). The findings of this research are provided in *Reuse of foundations for urban sites – A best practice handbook* (Butcher *et al*, 2006) and *Proceedings of the international conference on reuse of foundations for urban sites* (Butcher *et al*, 2006). See Appendix 1 for a summary of the RuFUS project.

1.2   Historical context

Britain has a rich and varied history; its cities have been continuously developed and remnants of this heritage are still present in the ground. A typical history for buildings on a central London street is shown in Figure 1.3.

(a) 1860s, Shallow foundations   (b) 1950s, Deep foundations   (c) 2002, Reused foundations (refurbishment)

Figure 1.3   *Typical history of buildings in Fitzroy St, London W1, culminating in Arup's new headquarters ((a) after Bailey, 1981)*

Reusing foundations used to be the norm rather than the exception. Important structures, such as cathedrals and castles, tended to be rebuilt on the foundations of their predecessors. In Elizabethan times in London (1558 to 1603), in an attempt to curb urban sprawl, new building was only allowed if it were raised "on old foundations". On the day that the Great Fire of 1666 was finally extinguished, King Charles II was informed that:

> *Some persons are already about to erect houses again in the City of London upon their old foundations (Ackroyd, 2001).*

However, as buildings have become bigger and expectations have increased, acceptance of cracks or other damage in structures by building occupants has decreased. As structural materials, finishes and services have become less tolerant to differential settlements, and methods for calculating foundation size have become more reliable, it has become standard to install new foundations to prevent both aesthetic and structural damage.

Foundation loads have continued to increase as developers seek to fill sites with the tallest buildings and lengthen floor spans, providing flexible space and increasing the available lettable area. This requirement for increased foundation capacity has led to bored piling and diaphragm walling techniques becoming common, largely since the 1950s.

As buildings have become bigger, the volume of underground services to serve the buildings has increased leading to a complex network of services under the footways and roads of most major cities. Another major user of underground space which has been developing since the early 1900s is the proliferation of underground transport systems eg in London, Newcastle, Liverpool and Glasgow. New schemes are continually planned, eg the Crossrail underground rail scheme beneath London.

All this history contributes to the accumulation of potential obstructions for new foundations, but the volume of ground available is rapidly diminishing.

An emphasis on preserving archaeological remains *in situ* also limits the space available for new foundations. This is another major driver for foundation reuse (see Section 1.4).

In recent times, foundations have been re-engineered and successfully reused on projects, for example railway bridges and on several major buildings including Empress State Building and Thames Court (see Appendix 2).

1.3 Foundation options for new developments

Several foundation options are available for new developments on sites with existing deep foundations which clash with new foundation locations, as shown in Figure 1.4.

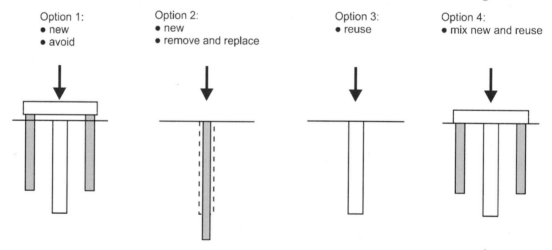

Figure 1.4   *Options for new piled foundations*

Option 1 – Install new foundations avoiding old foundations: Currently the most common strategy as it is perceived to be the easiest solution. As existing piles often occupy the prime locations on the site, new piles may need to be squeezed between these existing piles or installed away from the column positions. This can lead to large transfer structures between columns and foundations and requirements for higher capacity piles which will increase the cost and programme time, and may also affect the structural and architectural solutions.

Where avoidance of existing foundations is feasible, this option allows the redevelopment to proceed but adds to the obstructions in the ground, causing potential problems for future development. As the number of old piles builds up across a site the option to avoid them will become increasingly complex and expensive. On sites where archaeology is present, repeated new piling operations will be significantly damaging. So it is recommended that if any new foundations are to be installed, their future reuse is considered throughout the project (see Chapter 5).

Where an existing pile foundation is adjacent to a proposed pile location, the presence of the existing foundations may impede the construction of the new pile foundations potentially leading to a defective pile. The presence of obstructions, such as old foundations, near to the location of new piles should be considered and assessed as part of the pile design.

Option 2 – Install new foundations removing any old foundations: Where existing foundations or deep obstructions are on the column grid for a new building, removal is possible, even for deep piles, but the cost is high and there are programme implications. An example use of this option is where the old foundations were installed through archaeological remains and it has been determined that further piling on the site is undesirable. The new foundations would be constrained to the same locations and care would have to be taken during removal of the foundations to minimise damage to the *in situ* archaeology.

Removal of the pile, if fully reinforced, may be achieved by overcoring and progressive removal of pile lengths from the ground. This involves cutting out a circle of soil around the perimeter of the pile to separate the pile from the surrounding soil mass and allow it to be lifted out of the ground. Piles that are not fully reinforced can be problematic to remove by this technique, as the pile is likely to break and not lift out of the cored hole in a single piece. To overcome this, the pile can be broken down by rock boring equipment which will break the pile up into small pieces and brought to the surface. If insufficient records exist on the pile construction, then removal will be a major risk for the expected project cost and programme.

The process of removing the existing foundation may cause disturbance of the adjacent ground. The reduction in the capacity of new piled foundation at these locations due to softening or relaxation of the ground caused by this disturbance will need to be considered and assessed as part of the pile design.

As for Option 1, it is recommended that if any new foundations are to be installed, their future reuse is considered throughout the project (see Chapter 5).

Option 3 – Reuse the old foundations: If foundations are to be reused, effort will be needed to research and verify the existing foundation information. The associated costs may be disproportionately high for smaller developments. However, experience has shown that the reuse of existing foundations is currently viable where:

- piled foundations already exist and the site is further constrained by the presence of even earlier piles, archaeology or other underground structures eg tunnels

- the new building optimum column grid matches the location of the previous foundations

- information on the existing foundations is easily available and sufficient to provide confidence in their reuse.

This CIRIA guide sets out the background and key issues relating to this foundation option.

Option 4 – Install new foundations to supplement old foundations (mixed foundations): New foundations can be installed to supplement the capacity of existing foundations which are being considered for reuse. This hybrid solution can be used where the new loads are higher than those previously applied or if there is uncertainty over the original pile capacities or the quality of the piles. It can also be used where some of the column locations do not correspond to the existing pile layout.

It is advised that if this foundation solution is to be progressed, the geotechnical and structural engineers work together to identify and reduce the effects of any differential settlement on the new building.

Most of the pertinent issues for full foundation reuse are also relevant to mixed foundations. This CIRIA guide sets out the background and key issues relating to this foundation option.

Foundation options for sites with existing shallow foundations

Similar options are available for existing shallow foundations but due to the relative ease of removing existing shallow foundations, the drivers for foundation reuse rather than replacement are significantly lessened. Unless there are particular site conditions favouring reuse, replacement is likely to be the more favourable foundation option.

1.4   Drivers for change

The principal drivers for foundation reuse are:

Current ground congestion: As previously described in Section 1.2, the ground beneath major cities is congested with abandoned piles, archaeology, tunnels and other constraints on locations for new piles, illustrated in Figure 1.5. The problem is intensified with every new foundation system installed.

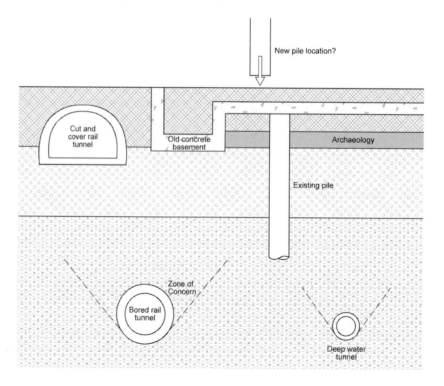

Figure 1.5   *Typical obstructions underground*

Figure 1.6 shows the how the congestion due to a set of existing deep foundations restricted the locations for a new set of foundations. To avoid the existing pile locations, various pile and pile cap configurations were required.

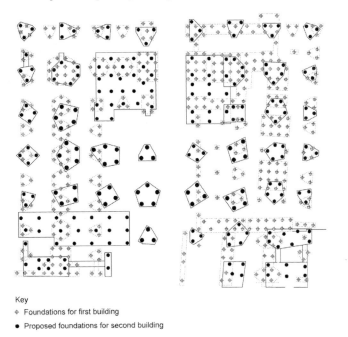

Key

⬦ Foundations for first building

● Proposed foundations for second building

Figure 1.6

Ground congestion from a first set of piles limiting locations for a second set of piled foundations (project in central London)

Archaeology: Archaeological deposits are present beneath most towns and cities in the UK, in many cases dating back at least 2000 years to the Roman period. Due to the depth at which many of these deposits are buried, previous development, particularly in the 1960s and 1970s, was less sensitive to their presence than is the case today. This has meant that archaeological deposits are preserved beneath and adjacent to site foundations. Currently, the preferred system of managing archaeological remains is to seek to preserve them *in situ*. In England, for example, this approach is outlined by The Government in PPG16 Planning Policy Guidance: *Archaeology and Planning* (Department of the Environment, 1990), which states that:

> *Where nationally important archaeological remains, whether scheduled or not, and their settings, are affected by proposed development there should be a presumption in favour of their physical preservation.*

Construction can be damaging to archaeological remains (Davis *et al*, 2004), and piling in particular (Biddle, 1994 and Nixon, 1998), as can be seen in Figure 1.7. When sites containing archaeology are re-developed, decisions about how preservation should be achieved may, on occasion, hinder or entirely impede the installation of new foundations. Negotiation with the planning authorities and their archaeologists may allow replacement of old foundations or some installation of new foundations subject to archaeological recording of the deposits affected by construction activity. The most appropriate method of reducing piling damage is to reuse existing foundations. Where new foundations are required, these should be constructed in such a way that they can be reused in the future.

Figure 1.7   *No 1 Poultry, City of London: Piling through archaeological deposits (courtesy Museum of London Archaeology Service)*

Minimising future ground congestion: Major cities are expanding and the number of new developments with deep foundations continues to increase. This growth also results from infrastructure improvements, which include underground construction. For example the expansion of London's underground transport network through the construction of Crossrail will add to the ground congestion. Thames Water's proposed Tideway tunnel scheme beneath the River Thanes will limit locations of foundations for new bridges over the river.

If the redevelopment of city centre sites continues with little consideration for what is happening below the ground, underground congestion will make city centre sites increasingly expensive to redevelop in the future. It shows foresight to leave some space for future foundations and not utilise all the available space unnecessarily. Consideration should be given to reusing foundations earlier than might be strictly necessary, to allow versatility for the future.

Potential cost and programme savings: The need to install a new substructure and set of piles for each new building is expensive and time consuming. The reuse of existing foundations presents an opportunity to reduce the piling requirements and in some cases may even remove piling from the project cost and programme. Potential cost and programme savings are explained further in Chapter 3.

Sustainability: There has become a greater appreciation of the need to protect scarce resources. This encourages reuse of all structural elements including foundations. The reuse of existing foundations and the design of today's foundations for future use need to be considered by project teams.

Most sustainability ranking systems reward sites that are close to existing infrastructure and offer good public transport facilities – city centre sites are considered more sustainable than building on greenfield sites. However, the construction of new buildings, with a new set of foundations, every 25–50 years, is not sustainable for the ground beneath cities. The option of foundation reuse should be considered seriously at an early stage in the redevelopment process to avoid the need to divert new development to greenfield sites.

Minimising energy consumption both during construction and operation of a new building is also a measure of sustainable redevelopment. Table 1.1 gives a relative comparison of the energy consumption for the four foundation options. Foundation installation is an energy intensive process that consumes large quantities of non-renewable materials – principally cement and stone aggregate. Reusing foundations should show a quantitative reduction in the use of these materials and reduce the "embodied energy" in a new development thereby improving its "carbon accountancy" balance. A new building, unencumbered by constraints is likely to operate most efficiently over its lifespan. Care should be taken to ensure that applying the constraint of an old foundation layout on the new building does not result in a less efficient building with operational energy requirements over the building lifespan that are disproportionate to the new material and installation energy benefits of foundation reuse.

Table 1.1   *Comparison of energy consumption for different foundation options*

Foundation option	New material inputs (piles and transfer structures) • concrete • reinforcement	Installation energy input • pile rig energy	Building operational energy usage • building running costs
1 Install new avoiding old foundations	High	High	Lowest (when new building is not constrained)
2 Install new and remove old	High	Very high	Low
3 Reuse old foundations	Low	Low	Medium to high (when building is constrained)
4 Install new to supplement old foundations	Medium	Medium	Low to medium

A further environmental benefit of foundation reuse is the reduction in breaking out and disposal of old substructures. This reduces the disruption to neighbours due to noisy breaking out activities and will also reduce the pollution and costs associated with transportation of the broken out material to appropriately licensed landfill sites.

Consideration of the city specific drivers for foundation reuse are shown through a case study for London (see Case study 1.1).

London is a major city now implementing pile reuse for many large developments. Considering why foundation reuse is becoming more important to this city should help to indicate which other cities will have a similar need in the near future:

Geology

A relatively soft underlying geology in London means that large buildings are often most efficiently supported on reasonably deep piles or under-reamed piles. A harder geology like the gneiss and schists that are beneath lower Manhattan Island in the centre of New York allows greater use of shallow foundations even for tall buildings, which are considerably easier to remove and replace when the building they support reaches the end of its life. Removing deep piles and under-reams in London is expensive, so they accumulate as successive generations of structures are built.

Other UK cities, eg Cardiff, have rock at shallower depth than London so piles are often shorter or have smaller diameters (and no under-reams), so there is currently more space for new foundations and in the future pile removal may be an easier option.

Land value

The centre of London has some of the most valuable land of any city in the world, providing space for large offices that house some of the most highly-paid employees. It is often considered cost-effective to replace buildings with more efficient ones more frequently. While the standard design life for a building is 50 years (BS EN1990:2002, Table 2.1), many buildings in London are demolished after just 20–25 years. This is more often due to changes in building usage or corporate identity, than to building performance or materials. As the use of deep pile foundations became widespread in the 1950s and 1960s, some building plots are now starting to have their third set of deep foundations installed.

As land values rise in other UK cities this factor will increasingly apply to them as well. Although land prices may not rise to the levels of those in London, if a building is replaced every 30–35 years due to changes in building usage, a greater need to reuse foundations is likely to occur over the next 10 years or so. If the replacement cycle is 50 years, ie the design life of the building, the problem will be deferred for another decade or so but will still be a future constraint for city development.

History

The UK has a rich heritage, and most cities contain archaeological deposits that add to the below ground construction constraints. London has been inhabited almost continuously since Roman times as one of Britain's major centres, and so contains a vast quantity of archaeology, much of which is of significance and worthy of long-term preservation.

A further historical factor affecting reuse is that the centre of many major UK cities, and especially London, were devastated by bombing in WWII, freeing up the central land for new buildings. Unlike structures which survived the bombing, these new developments do not tend to have historical significance and are easier to replace. By contrast, the heart of Paris was not fought over (thanks to General Dietrich von Choltitz who disobeyed several direct orders not to let the city fall "except lying in complete debris") and the city centre is full of buildings and façades worthy of protection.

These historical factors are particularly relevant to London, as although many of the UK's major cities were also devastated by WWII bombing, few of those also contain the same wealth of archaeological remains.

Planning laws

The UK is subject to tight and actively enforced planning laws that place a high value on heritage. British cities are more likely to be constrained by a need to protect archaeology than cities in other parts of the world where the need for development can outweigh protection of historical and archaeological features.

2   Requirements for any foundation

Before embarking on this review of the principal issues of foundation reuse, it is important to define the key functions of any foundation so that particular aspects of reuse can be seen in context. The key functions of foundations are reliability and safety, building performance, and durability.

Close collaboration between structural and geotechnical designers is fundamental if existing piles are to be incorporated as integral elements of the new structure. The role of both co-designers cannot be underestimated, as the structural requirements for the new building are paramount and will be based on a broad understanding of the sensitivity to settlement of all the building elements, including services and finishes. The reuse of old piles cannot be considered unless the structural criteria are clearly defined. Conversely, a structural team without knowledgeable geotechnical support may overlook important aspects of the reuse process such as the different settlement characteristics of new and reused foundations under loading (see Section 4.5). Many of the case studies given in Appendix 2 relate to the "re-engineering" of existing piled foundations to be used in conjunction with new piles and the management of different settlement characteristics.

2.1   Reliability and safety

It is a fundamental requirement of all foundations that they are reliably safe. Reused foundations need to have a comparable level of reliability to newly installed foundations. This is principally concerned with ensuring that the building will not collapse, even if overloaded by a possible margin.

The building developers are dependent on this reliability if they are to successfully and profitably lease the building for its useful design life, or sell it on. The potential costs of unreliable foundations are presented in a paper by Chapman and Marcetteau (2004), which shows that the economic consequences can be disproportionately severe if problems occur in the foundations.

The design factors of safety required by design codes allow for variability in design parameters and applied loads. Geotechnical factors of safety tend to be higher than purely structural factors because of the inherent uncertainty of heterogeneous ground, imperfect installation techniques and settlement limitations.

For a foundation system that has already been tested by the application of the first building load, the allowable factor of safety against failure may be lower than would be used for conventional foundations, provided that sufficient details of the existing foundations were known and the disposition of loads is similar. Eurocode 7 (BSI, 2004) potentially allows an overall geotechnical factor of safety lower than 2 for new foundations using more reliable installation methods and with more than two preliminary pile tests to validate pile capacity and settlement performance. If the application of the first building load could be proved to be the equivalent of a load test, then it could be argued that a similar value could serve as the minimum factor for reused foundations. However, this approach has not yet been used on a project and omits the fact that the first building load does not prove the factor of safety by the extent that a preliminary pile test would. In particular, it is difficult to assess the actual building loads carried by the first building. These are likely to be less than the design loading primarily due to the design assumptions of the magnitude and distribution of live loading in the building (see Section 4.6). For a future design approach to adopt a lower factor of safety for reused foundations than the original design case, there would need to be sufficient information on the existing piled foundations and the foundation performance would need to be considered explicitly (see Section 2.2).

Professional site supervision records from the installation of the existing foundations are helpful to ensure that construction was carried out in accordance with the design, and that any anomalies were recorded and checked to ensure that the ability of the structure to fulfil its requirements over the design life was not compromised. This gives the development team confidence that their foundation scheme is reliable. Construction records will also give future redevelopment teams confidence that the foundations to be reused were properly installed (see Sections 4.1 and 5.1)

It is normally easier to justify foundation reuse to support vertical compressive loads because they are reliant on parameters which can be predicted with more certainty, eg pile dimensions, geotechnical capacity and concrete durability. Reuse of piles to support tension loads, horizontal loads or applied moments can be more problematic and involve higher risk (see Figure 2.1). This is due to the reliance that needs to be placed on pile reinforcement details and lack of corrosion, which could be expected to vary across the foundation system. These may not be adequately recorded on drawings, and cannot easily be verified by site testing.

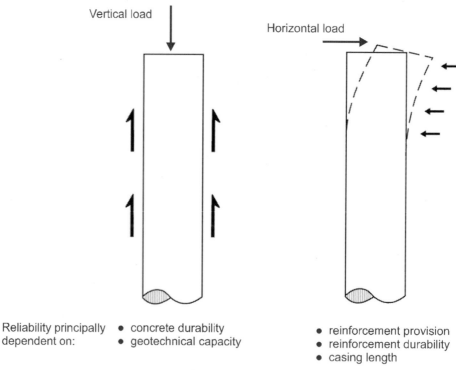

| Reliability principally dependent on: | • concrete durability
• geotechnical capacity | • reinforcement provision
• reinforcement durability
• casing length |

Figure 2.1   *Reliability of existing foundations*

2.2   Building performance

Another fundamental purpose of the foundation system is to provide acceptable settlement performance at or just above its working load for the duration of the building life. Acceptable performance is normally defined from consideration of differential settlements to avoid cracking in the structure, damage to architectural finishes, impairment to the operation of services or the function of the building. If the performance and function of any element of the building is sufficiently impaired, the building may become unusable.

Where reused foundations are mixed with new ones, consideration should be given to ensuring compatible settlement behaviour between the different types of foundations at working load. These aspects are discussed further in Chapter 4.

One significant step towards allowing foundations to be reused is to make the superstructure more tolerant of the potential for differential settlement between different columns and foundations. This can be accomplished by the provision of stiff shear walls to distribute loads between different foundations in case any part of the foundation system under-performs. This is particularly useful when there are doubts about the reliability or performance of the existing foundations.

2.3   Durability

Deep foundations in city centres are often bored piles and with foundations commonly consisting of two materials – concrete and the embedded reinforcing steel. Driven piles are seldom used in congested areas due to noise and vibration issues, but if used are generally steel sections or reinforced concrete. Timber piles, although not in current general use in Britain, were used in cities developed on estuaries in late Georgian and Victorian times, such as Belfast City Centre and London's Docklands (timber piles are still used in some places such as Scandinavia and New Zealand). Each pile type needs special consideration of the specific processes that could cause deterioration.

If foundations are to be reused, their structural capacity needs to be adequate for the proposed new design life.

When considering the reuse of foundations it is important to investigate the extent of any material deterioration and estimate the effects of any damage on the future performance of the system. This involves ensuring that:

● the foundations are sufficiently undamaged at the time of inspection so that future deterioration processes will not inhibit their function

● there is no evidence of the deterioration processes having already started at a rate that gives cause for concern.

Field investigations, including those relating to durability are identified further in Section 4.4.

A key issue in assessing old foundation materials is the proposed new design life that is to be allocated to them. Materials used in pile construction are chosen for their durability and it is often the case that at the end of their first life, there is little evidence of deterioration. There is normally little difference in the quality of piles specified for buildings and for civil engineering despite having normal design life of 50 years and 120 years respectively in accordance with the British National Annex for BS EN 1990:2002 (BSI: 2002). For buildings a minimum new life of 50 years is needed unless otherwise agreed.

3   How can foundations be reused?

The decision to reuse foundations will be based on an assessment of each individual site. In some cases, where low loads are to be applied to high capacity foundations, the additional risks will be low. In other cases, where old foundations of doubtful provenance are to be heavily loaded, the additional risks associated with foundation reuse will be higher. At the concept design stage, costs and programme implications for different foundation options need to be assessed. At the detailed design stage, a decision has to be made on what foundation option is considered the most appropriate for the redevelopment. Even if a developer is amenable to the concept of reusing old foundations at the start of the project, findings from an investigation into the existing foundations may make the option look increasingly risky and subsequently less attractive. There needs to be decision points at which the whole development team can affirm their continued acceptance of the concept of reusing old foundations, or reverting to the more conventional installation of new ones. In all cases where foundations are being reused, a pre-requisite is a careful desk study and the integrated skills of both structural and geotechnical engineers.

If foundations are to be reused, then it is important that the following factors are considered.

3.1   Foundation information

Information on the existing foundations (type, location, diameter, length, reinforcement) is essential for foundation reuse. This information should be obtained as part of the desk study process and confirmed by field investigations especially where any doubt as to the quality or accuracy of the information exists. Often the original design and construction companies may no longer exist. Obtaining original records can be problematic and detailed data may not be available (see Chapter 4).

3.2   Compatibility

It is seldom possible for the layout of a new building to be constrained to an existing column grid and also to be economically able to fulfil its new function. Where the new building layout is mostly compatible with the old column grid, it may be possible to reuse the existing piles supplemented by a relatively small number of additional new piles. However, where the new building is generally incompatible with the old column grid and where new piles are not feasible due to other obstructions in the ground, a raft or large transfer structure will be required to spread the load to the available foundations.

Figure 3.1 compares a proposed column layout with the existing foundation layout for a project where foundation reuse had been considered desirable. In this case the new columns and the old foundation only coincide at about 10 per cent of the locations and foundation reuse was considered inappropriate for the development.

Figure 3.1
Incompatibilities of proposed and existing column locations (project in central London)

3.3 Foundation capacity

The reuse of existing piles is already accepted practice for nearly all refurbishment projects. For these projects the configuration of the new loading is similar to the original design loading, and foundations can receive an increase in loading. BCA/BSRIA (1999) state:

> *Extra load capacity is generally built into the foundations of older buildings and in the 1970s building control would commonly allow an additional 10 per cent of the total building loading to be added to the foundations at a later date, provided the building was sound and no settlement had occurred.*

As a result of planning restrictions, many city centre redevelopments are constrained to a similar height as the previous building but often spans are increased. This results in the need for a smaller number of more heavily loaded columns than before. Even if the new columns do coincide with the old foundations, those foundations are unlikely to have enough capacity for the new loads, without some additional foundations.

The results of limited load testing indicate that in most circumstances foundation capacity increases with time (Whitaker and Cook, 1966, Wardle *et al*, 1992, Chow *et al*, 1997 and Chapman *et al*, 2001), however this apparent spare capacity is often difficult to justify to the regulatory approval bodies. So the pile reuse capacities are often restricted to the design loads of the previous building (see Section 4.6).

An important check when reusing foundations is that the original foundations performed adequately. If there is evidence of poor foundation performance, eg records of structural damage or repair to the first building over its lifetime, the viability of reusing the foundations should be reconsidered more carefully, and it may be necessary to reduce the design reuse load capacity or even discontinue consideration of foundation reuse (see Chapter 4).

3.4 Design codes and regulatory approval

Reuse of foundations is not explicitly addressed by current British or European standards (for instance BS 8004 and Eurocode 7), and both expect a new set of foundations will be installed for each new building. However, design codes evolve and future revisions may provide explicit advice on re-engineering of existing foundations. *The Building Code of the City of New York* (2003) does specifically address the "*Use of existing piles at demolished structures with requirements for assessing the load bearing capacity for reuse*" (Clause 27-690).

Foundation design incorporating reused foundations may not fully comply with current codes, particularly for materials, so a pragmatic approach is required. As an example, the proportions of cement needed to resist sulfate attack increase with each evolution of the published advice (BRE Digest 250, BRE Digest 363, BRE Special Digest 1, 2001 and 2005). Yet, in most circumstances, concrete from a previous era that has survived buried in a ground environment without any significant deterioration should be able to have another design life safely assigned, irrespective of current guidelines (see Chapter 4).

To adopt foundation reuse, the developers will need to explicitly agree that they are prepared to accept a building that may not fully comply with current standards. This acceptance will also need to be agreed with any financial backers, future insurers and also regulatory bodies such as the Building Control Office.

3.5   Identification of risk

Foundations need to be economical and reliable. Developers are invariably risk-adverse and are normally prepared to pay a premium to avoid uncertainty. With the current perception of the risks associated with foundation reuse, developers often prefer to pay more to install new foundations, including removal of any existing pile obstructions, than to seriously consider foundation reuse.

However, the perceived risk of pile reuse is often greater than the actual risk.

The best way to manage risk is to identify the hazards from the start of the project and mitigate them. Identified hazards can be taken into account by the building design team. The structural and geotechnical designs should be considered in combination in order to provide an appropriate level of overall foundation robustness to deal with the identified risks.

3.6   Cost

Cost is a major issue to the developer. As foundations cannot be seen, they are often an element of structure for which the developer will wish to minimise costs.

The cost of the various foundation system options depends on several factors including:

- congestion of the ground beneath the site in terms of obstructions, number of old foundations, archaeology etc
- the applied loads and ground conditions
- information available on previously installed foundations.

The total cost of the foundation system will comprise:

- direct costs eg material and installation costs
- risk costs eg programme delays, repair costs.

Foundation systems need to be economical and reliable. Striking a balance between economy and reliability is important for all parts of the building, including foundations. A conservative solution could benefit from a small investment in better design or investigation resulting in significant cost savings and a possible reduction in risk. Over-optimistic designs may result in underperformance of the structure and unforeseen costs such as repairs.

Table 3.1 shows an indicative comparison of cost elements for the four foundation options identified. For new piles, the table considers bored piles as opposed to driven piles as driven piles are seldom used in city centre locations where foundation reuse is most likely to be adopted.

Table 3.1

Relative cost elements for foundation options (for bored piles)

Foundation option	Material costs	Disposal cost	Design costs	Investigation costs	Insurance costs
Option 1 Install new piles avoiding old foundations	High	High	Medium	Medium	Low
Option 2 Install new piles, removing old foundations	High	Very high	Medium	Medium	Low
Option 3 Reuse all existing piles	Low *	Nil	High	High	Potentially higher
Option 4 Install new piles to supplement existing piles	Medium	Low	High	High	Potentially higher

* may need transfer structure

Traditionally, it has been assumed that the cost of a building was limited to its construction cost, and that the interest of the developer was best served by delivering the cheapest constructed solution. However, there is an emerging awareness that the lowest construction cost solution may not provide the most economically favourable overall outcome for the developer. The majority of commercial developers are interested in deriving maximum income from their development. Thus other costs and benefits have also to be considered, including early opening of the development and lower maintenance or running costs.

Investing to reduce risks can be an effective way of reducing the construction cost, programme duration or whole life cost, or achieving other developer requirements.

3.7 Programme

Minimising the duration of the construction programme for a redevelopment can be key to minimising the overall redevelopment costs. Successful implementation of foundation reuse has the potential to reduce the construction programme and reduce redevelopment costs.

If a reliable platform was available from the previous building on which a new superstructure could be founded, the construction could be completed significantly faster. This principle was implemented for the Phase 1 buildings at the western end of Canary Wharf, where reinforced concrete platforms were built over the docks onto which the superstructures were mounted.

Additionally, the riskier operations involving work in the ground would be eliminated giving greater security of programme. This would also result in buildings being unoccupied for a shorter period of time, which should reduce the loss of revenue during redevelopment and reduce the developer's exposure to interest charges on money borrowed to finance the redevelopment.

Table 3.2 is an idealised programme indicating the potential benefit in reusing old piles and a further benefit from reusing a preconstructed platform over a traditional construction programme involving the installation of a new set of piled foundations. The programme assumes that the old foundations are suitable for reuse, with good information available on them from the outset. A preconstructed platform would be designed to support a variety of load combinations for successive superstructures. This allows a new building to be constructed on the existing foundation system without need for strengthening or augmentation (see Section 5.3)

Table 3.2   *Typical construction programme timings*

	Typical reconstruction programme (years, months)	Reconstruction with reuse of old piles (years, months)	Reconstruction programme on preconstructed platform (years, months)
Ground investigation	0,3	0,3	0,3
Planning	0,6	0,6	0,6
Demolish	0,6	0,6	0,4
Remove obstructions	0,6	0,1	0,1
Investigate existing foundations	0,0	0,2	0,1
Piling and embedded walling	0,6	0,3	0,1
Excavate and build basement	0,5	0,3	0,1
Build superstructure	0,9	0,9	0,9
Fit-out	0,6	0,6	0,6
Total	3,11	3,3	2,8

If foundation reuse is desirable, considerable field investigations of the existing piles may be required, especially if the existing records are poor. Some of this investigation work is likely to take place when demolition of the existing building is in progress or complete, allowing access to the foundations. This places uncertainty on the project's critical path. If the available information cannot be verified and, as a result, the foundations cannot be reused then new foundations will be needed which will have an impact on both the construction costs and programme.

Figure 3.2 compares the likelihood of programme duration for a typical construction programme with foundation reuse. Where reuse is validated by an investigation at demolition or construction stage, foundation reuse shows programme benefits. However, the delays caused by late discovery that foundation reuse is not feasible can be clearly seen. If the need for new foundations is identified at this late stage there is a risk that the overall programme will be longer.

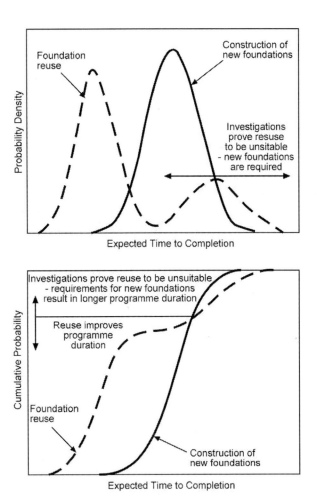

Figure 3.2   *Comparative probability of programme completion*

When considering the adoption of foundation reuse for development projects, the risk of programme delay should be assessed. For the idealised case study shown in Figure 3.2 there is approximately a 75–80 per cent probability that foundation reuse will result in a shorter construction programme. In practice the risk of programme delays where reuse is being implemented will depend on the quantity and quality of records available to the design team at the early stages of the redevelopment project. The developer's willingness to risk delay for a potentially shorter overall construction programme will depend on his financial ability to accept uncertainty.

3.8   Risk and liability

The risks associated with pile reuse are analogous to the risks associated with unforeseen ground conditions, or those encountered during refurbishment of an existing building:

- uncertainties regarding what cannot be seen prior to construction

- with attendant risk of remedial measures being required during construction leading to cost and time overruns, and possible compromise of the developer's aspirations for the building

- in extreme situations, abandonment of the scheme or even building failure.

Experienced construction professionals should be able to address and manage these risks on their developer's behalf. In particular, they should be able to quantify and explain the risks to dispel misconceptions and they should be able to advise on risk mitigation strategies. As with ground risk, early investment in investigation and feasibility assessment will pay dividends later.

The developer's engineers need to ascertain the developer's priorities and his appetite for such risks at the outset. The developer should be made aware of his potential liabilities as well as the potential liabilities of those he employs as advisers or constructors:

- the original designers/constructors (if they can be found) are unlikely to offer any warranty in relation to the as-built state of the original piles. Any information they offer is likely to be provided on an "as is" basis, with no warranty

- the developer's design team will need to identify the level of information and form a view as to its accuracy. It is important to note that if the design team makes engineering judgements which subsequently turn out to be incorrect (eg over-estimating the capacity of existing piles), they will only be liable if the developer can show those judgements were negligent

- the result will be that unless the developer has passed the risk associated with existing pile capacity to the contractor, the developer will retain it

- seeking to pass the risk to the contractor via a "fitness for purpose" building contract may appear to offer comfort but there is a premium to be paid and if the risk materialises, it is unlikely that the developer will escape loss entirely. In some situations, for example contractor insolvency, the risk will revert to the developer

- similarly relying on collateral warranties to give building users a right to seek redress is not necessarily an effective means of transferring the risk.

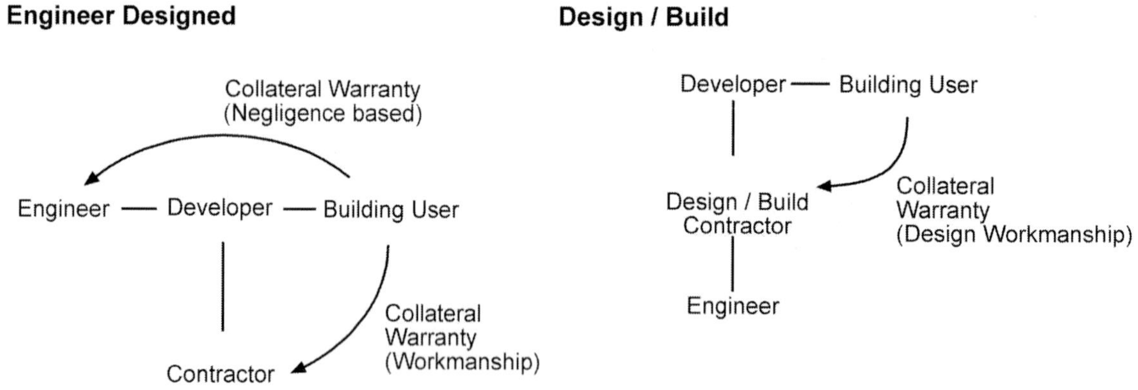

Engineer Designed   **Design / Build**

Figure 3.3 illustrates the typical UK construction contract structures.

Figure 3.3   *Typical UK construction contract structures*

The decision on whether or not to reuse existing piles will usually be made after a desk study and preliminary investigations have been completed (see Chapter 4). It is recommended that a report is prepared setting out the options for the developer, explaining the risks and seeking to evaluate the impact in terms of time, cost and quality of the risk materialising. Well advised professionals will ensure they have their developer's informed consent if they adopt the reuse of existing foundations as part of their overall foundation strategy.

3.9 Addressing building user's concerns – latent defects insurance

A project's commercial viability is often driven by the letting market. The commercial developer will often aim to sell on, or let on a long lease, the completed building. Tenants/purchasers may have negative perceptions about a building founded on reused piles and this may impact on the developers' return. A solution is Latent Defects Insurance (LDI). This covers repair costs (but not necessarily consequential losses) following practical completion. It is generally available in the UK and has been adopted for buildings incorporating reused foundations in institutionally funded developments. It applies for a fixed term – typically 12 years and is similar to the Decennial Insurance that operates in France and in other countries whose legal systems derive from the French legal system. LDI providers normally require auditing of the design and construction by another firm of consulting engineers – this extra rigour of itself will help to reduce the likelihood of defects.

Where the developer is planning to occupy the building for his own use, he may be more prepared to consider reused foundations as he may be less concerned about perceived risks impacting on marketability.

4   Technical considerations for reuse of old foundations

The process of verifying foundations as adequate for reuse depends on many factors, in particular on whether good information exists from their installation. The decision to proceed with reuse needs to be progressively reviewed as information becomes available (see Figure 4.1).

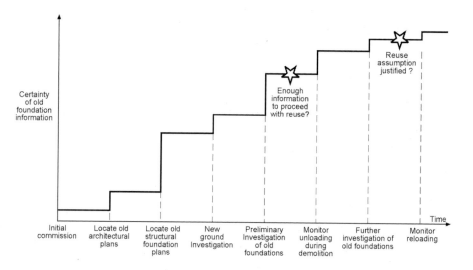

Figure 4.1   *Information to be collected as part of the reuse design process*

Early activity in these information gathering and exploratory stages will be beneficial to the project by providing early indications and reassurances as to the feasibility of foundation reuse. As described in Chapter 3, late identification of problems related to foundation reuse can have a significant negative effect on the project programme and construction cost.

4.1   Desk study

A desk study is an integral part of any development design process. Where foundation reuse is a potential option, the importance of a thorough desk study cannot be overstated.

In addition to more standard information such as site history, archaeological and contamination potential, the desk study should collate all of the available information on the existing building, foundation system and transfer mechanisms. Information collected should include as-built information including any drawings and non-conformances from the installation of the existing foundations. This will help manage the risk of poor construction quality of the existing piles which will increase confidence in pile reuse for the project.

Design and construction records may be stored in a variety of locations. Possible storage locations include:

- the building control offices of the local authority. Note that they may not have preserved records due to lack of space, and increasingly will not allow access without written authorisation from the building owner

- the original designers: architect or structural engineer or the original foundation contractor. Practices may have merged or gone out of business. A fee may be charged to cover the costs of locating and copying the documents and this may be set at a level commensurate with the perceived value of the records to the new developer. Often the available records are very limited

- for structures constructed since the introduction of the Construction (Design and Management) Regulations (CDM) in 1994, building owners should have a set of records for the building construction – however few of these buildings have been demolished at the time of writing this publication.

An example of good luck in locating old records is given by St John *et al* (2000). They recount how a time capsule, buried on 21 September 1961 when a building was being built, was found in 2000 in a footing being broken out in a trial pit to appraise the extent of the existing foundations. Amongst other artefacts in the capsule was a drawing showing the foundation layout. Few other developments will have such good fortune.

The information available to the designer will dictate the extent and efficiency of any pile reuse. Table 4.1 summarises information that should be collected. Where possible any information to be relied on for design should be verified during the site investigation.

Table 4.1   *Information to be collected for a desk study*

Information	Desk study - possible information sources	Field investigation - methods of collecting/verifying information
Pile diameter	Design drawings and specification confirmed by as-built drawings and schedules Calculations – design loading/load take-down from previous building	Intrusive investigation and testing Non-destructive testing (NDT) Visual investigation (see Section 4.4)
Pile length		
Pile position		
Reinforcement details		
Grade of concrete		
Permanent liners		
Construction quality	As-built records including pile testing records	
Damage/deterioration over first design life	Design information should be treated with caution as it is not known what errors or omissions may exist.	
Damage during demolition		

4.2   Assessing current conditions

It is important to confirm that the first building performed satisfactorily. The building should be checked for evidence of settlement (for instance by observation of the façade for evidence of repairs, by checking records for any remedial works, or by levelling a building course – eg at the damp proof course level around the building perimeter). For internal surveys it should be borne in mind that buildings that have suffered from settlement may have had the floor refinished to provide a level surface and so a survey of the ceiling may provide a better indication of movement.

The history of loading of the building should then be assessed. Is it currently fulfilling the same use as that for which it was originally built? Over its history, is it likely to have been loaded to a greater extent than at present? Consulting old trade directories may give useful information on previous building uses which will provide an indication of historical loading conditions.

4.3   Monitoring to verify performance

Monitoring existing operational buildings can provide useful data for design teams considering the reuse of old foundations. Various types of monitoring can be used to verify different aspects of the structure's behaviour.

Monitoring and assessment for cause of defects: Structural defects which are suspected as having been caused by a foundation failure should have a profound effect on the confidence of placing a new building on the old foundations. As only a small proportion of reused foundations can ever be comprehensively checked, any observations of structural damage in a first building cast doubt on the reliability of the rest of the foundation system. A greater extent of testing than normal would be required to provide confidence before reusing the foundations from a building where the first set of foundations did not appear to have functioned adequately. Otherwise only a smaller proportion of load can be safely applied.

For buildings where foundation under-performance is suspected, and where the foundations are proposed for reuse, it is vital to correctly diagnose the cause of any previous structural damage to the first building, and monitoring may be a useful element in determining the diagnosis. As normal operational structural movements tend to be small (and possibly immeasurable), they seldom influence the viability of foundation reuse, unless they show evidence of unexpected foundation or subsoil movements.

Monitoring during unloading/demolition of first building: Theoretically, the data will provide an indication of foundation performance on unloading, and subsequently give an indication of performance for reloading. Monitoring during demolition may provide very useful data for assessing the viability of the old foundations for reuse.

Load cells are typically used for monitoring loads in the columns or piles and a variety of targets can be used for monitoring movements, both vertical and horizontal. Depending on the chosen instrumentation, monitoring may be carried out by manual or automatic methods. The choice between the methods will depend on their relative costs, which will depend on the frequency of readings, the duration over which they are to be taken and the difficulty in achieving access.

There are a number of issues that may affect the quality of information obtained by this method:

- actual load at each column and hence into each foundation is unknown, so it is difficult to calculate the stiffness of individual foundations

- even more so than for construction, accurate monitoring during demolition can often be impaired by monitoring points being knocked or damaged or indeed demolished

- monitoring points may need to be relocated due to demolition activities such as transferring level stations to newly exposed pile heads as the columns and slabs are removed. This may provide a source of error in the monitoring results

- often the unload displacements will be relatively small and will take place over a longer timescale than just demolition, so the monitoring may provide little meaningful data to the team considering foundation reuse

- the relatively small magnitude of the movements put them at risk of becoming masked by general surveying errors or errors due to lost or transferred level stations.

Presently the problems with this approach mean that it is seldom used. However, as instrumentation becomes accurate, cheap and ubiquitous, it may provide great reassurance to project teams considering foundation reuse in the future (see Section 5.2).

4.4　Field investigations

Need for investigation

At present it is unlikely that sufficient records exist for previous foundations to allow the design for reuse to be carried out without any field investigations. Information from the original design stage of the existing foundations should always be used with caution as an indication of the final design and geometry of the as-built foundations. To ascertain the "as-now" properties of the foundations, available information may need to be supplemented or verified by some form of intrusive investigation. Care should be taken during the investigation not to unacceptably impair the foundations under study for their future life. For sites where archaeological remains exist, care should be taken during the field investigation to avoid damaging them or triggering a deterioration process.

The field investigations identified below are orientated towards reinforced concrete bored piles. In addition to confirming pile location and geometry, the purpose of these investigations is to verify that the pile material is not deteriorating unacceptably and that the pile element has not become unacceptably damaged or impaired at some stage over its life.

Types of investigation

The scope of the investigation is likely to involve breaking down pile caps on a representative sample of piles to expose the pile heads and allow pile dimensions and reinforcement details to be ascertained. *In situ* testing and sampling for laboratory testing can also be carried out to ascertain properties and condition of the pile concrete. All investigation techniques have advantages and limitations so it is recommended that specialist advice is sought at the planning stage and that the specific aims of the investigation should be clear. Checks for the more important parameters are shown in Table 4.2.

Table 4.2   *Hierarchy of pile examination for reuse*

Relative importance	Factor	Estimation test	Parameters proven
1	Location	Positional survey Visual examination of head NDT	Pile is present Pile location Approximate pile diameter Typical materials
2	Length	Transient dynamic response (soft hammer) test Geophysics Coring	May prove length May find defects
3	Material quality/ integrity	Sampling and laboratory material testing *In situ* material testing	Confirm material properties Show material degradation
4	Geotechnical capacity	Load test	Geotechnical capacity (to failure?) Pile integrity

Tests to assess the material quality and to aid in the assessment of pile deterioration are summarised in Table 4.3 and shown graphically in Figure 4.2. When planning investigation works, consideration should be given to minimising damage to the pile and changing the environment for the foundation. Exposing concrete or introducing new backfill material introduces oxygen into the soil which, if the soil contains sulfides (eg pyrites) may also provide a new source of sulfates through oxidation (Building Research Establishment, 2005). This may lead to sulfate attack in the future due to the newly elevated sulfate levels even if there was no evidence in its current state to suggest a potential problem.

Table 4.3   *Types of test for pile deterioration*

Material	Type of deterioration	Testing
Concrete	Sulfate attack Acid attack	Visual inspection Petrographic analysis Chemical testing Carbonation depth (if exposed to air)
Reinforcing steel	Corrosion	Exposure for visual examination Chloride ion content profile Cover to reinforcement
Timber	Rot Boring insect	Visual inspection

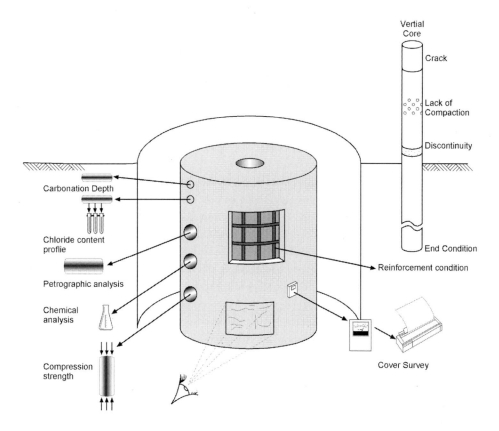

Figure 4.2   *Testing for pile deterioration*

Scope of investigation

The extent of investigation carried out should be agreed by all parties, including the risks inherent in carrying out too little or the costs of doing too much. An extremely costly field investigation would be to physically examine every old foundation element to verify that they are all adequate. This would have two principal disadvantages:

1   It could compromise all the existing piles, depending on the type of investigation carried out – the process of drilling holes to take samples and exposing the piles could lead to induced damage and increased deterioration.

2   It would be very costly and time-consuming – on a site where every pile has to be individually examined after demolition and prior to reuse, the exercise would add considerably to cost and programme, almost certainly making foundation reuse uneconomic.

Conversely, each pile not checked carries a theoretical risk, which may be significant depending on the relative magnitude of the imposed load and the reliance being placed on the individual pile. The balance on the extent of testing will depend on the perceived reliability of extrapolating from the available records and the levels of reloading which is proposed for the foundations (see Section 4.6).

The aggressivity of the ground conditions (eg sulfate level, pH, groundwater mobility), and whether it varies significantly across the site, can be found from the desk study, other prior experience and/or results of a new site investigation. This information can then be used to decide where it is most appropriate to select piles for examination, and how many. In terms of pile performance, correctly supplied concrete is not generally a highly variable material. So it is reasonable, in most cases, to assume that results from a small number of piles can be sufficiently representative provided there is not large variation in the potential aggressivity across a site. Where significant variation in ground conditions is expected, it may be prudent to examine the piles in the most aggressive areas first to confirm if the reuse strategy is viable.

Interpretation and implications of defects within and deterioration of existing concrete

The best guide to future performance of concrete in particular ground conditions is its past performance, particularly where the environment is unlikely to change significantly. If the testing described above shows little or no deterioration after a period of several decades then, without any significant change in ground conditions, it is highly likely that no serious deterioration will occur within the normal design life of a new structure. In assessing the likely future conditions, particular attention should be paid to the possibility of increased water mobility due, for example, to rising groundwater levels which could change the rate of supply of aggressive ions to the concrete surface from the surrounding ground. The most common defects and forms of deterioration are considered below:

- chlorides – the most likely sources of chloride are from sea water or brackish water or from run-off from roads where de-icing salts have been applied. For high strength concrete in the 1960s, calcium chloride was added as an admixture for strength acceleration. However there was less of a perceived need for this in foundations and chlorides are unlikely to have been added to a foundation concrete mix. Chlorides pose a risk of corrosion to reinforcement but only if sufficient oxygen and moisture are present and the chloride level at the reinforcement is greater than the "threshold value" for corrosion

- carbonation – carbonation in buried concrete is usually very low unless there has been exposure to air. Carbonation is only a risk where it reaches the depth of the reinforcement within the required service life and where sufficient oxygen is available to support steel corrosion.

These can be exacerbated by construction problems or defects in the completed pile section:

- poorly compacted concrete is only likely to be a durability problem if aggressive ground conditions exist and perhaps for reinforcement corrosion close to the surface where sufficient access to oxygen is available and the poor quality concrete has allowed carbonation or ingress of chlorides to the depth of the reinforcement. The same principles are generally applicable to large cracks

- surface deterioration and loss of surface integrity, such has been observed with the thaumasite form of sulphate attack (TSA), may be significant through loss of load carrying cross-section particularly for friction piles. However this is likely to be rare except in very aggressive conditions where concrete of insufficient resistance has been used.

Petrographic indications of internal chemical attack of the concrete material such as alkali-silica reaction, or the presence of sulfate reaction products such as ettringite or thaumasite should be treated with caution as these are not necessarily indicative of significant deterioration.

Where material deterioration issues appear to be significant, advice should be sought from a materials expert.

4.5   Foundation performance

The prediction of a foundation's performance on reuse requires an understanding of the soil structure interaction during the different phases of construction, demolition and redevelopment. An example of a pile's performance over two building lives is shown in Figure 4.3.

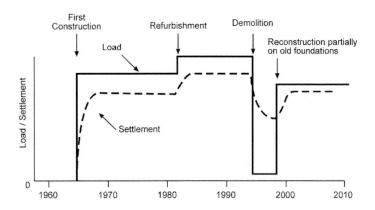

Figure 4.3   *Foundation performance with time*

The required performance from reused foundations depends on a number of factors, all of which relate to the sensitivity of the new building to settlement, and to differential settlement in particular.

The performance of old piles will tend to be stiff up to the load to which they were previously loaded. Loading beyond the previous maximum load applied will be less stiff and it may be more difficult to limit settlements between adjacent piles. It is suggested that new loads should not exceed those previously applied, unless differential settlements are calculated accurately and explicitly. Wherever old and new piles are being mixed in a new foundation system, settlements need to be calculated and the differential settlements communicated to the structural engineer for the design of the pile caps and structures linking the old and new piles.

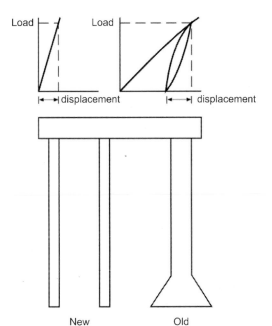

Figure 4.4

Relative settlement behaviour of old under-ream piles and new straight-shafted piles

As an example, when new straight-shafted piles are mixed with old under-ream piles with different settlement characteristics, the under-ream piles will be much stiffer on reloading and so more compatible with first time loading of long straight piles, see Figure 4.4. The reverse situation, new under-ream piles mixed with old straight-shafted piles, are unlikely to be compatible.

4.6   Assigning a new design/working capacity

To assist in the consideration of reusing foundations a parameter relating the load capacity required for the re-engineered foundation to the previous foundation capacity is helpful (Chapman *et al*, 2002 and Chow *et al*, 2002). This allows the load demand at each foundation point to be assessed.

$$R = \frac{\text{new foundation load demand}}{\text{old foundation working capacity}} \qquad \text{essentially the ratio between the new required working load and the old one.}$$

The capacity of the existing foundation at working load could be one of the following:

- *original calculations* (R_o) – technology and knowledge develop over time and the theories previously used could now be considered inappropriate for the ground conditions. Original design calculation should be checked for any errors

- *current theory and approach* (R_c) – this value can provide information on the expected margin against failure. Sufficient information on the geometry of the foundation needs to be available and requires verification, or the calculated capacity could be misleading

- *load takedown* (R_t) – a load take-down for the previous building should be undertaken based on the building dimensions and use, and is not reliant on foundation information. This assessment of the loading from the previous building in each foundation element should be based on past proven loads (from an assessment of real building usage over its life), not on previously stated design values

- *foundation test* (R_p) – load tests can be carried out, however, this has rarely been undertaken in the past.

The original calculations may give a lower capacity than those based on current theory and approach due to improved understanding of ground performance, validated by a database of static pile load test results. Piles constructed prior to 1980 were constructed more slowly than today and may have had lower capacities. Tripod bored piles in particular were built slowly and sometimes with greater seepages than that of modern pile-forming equipment. An illustration of the inter-relationship of the various capacities is shown in Figure 4.5.

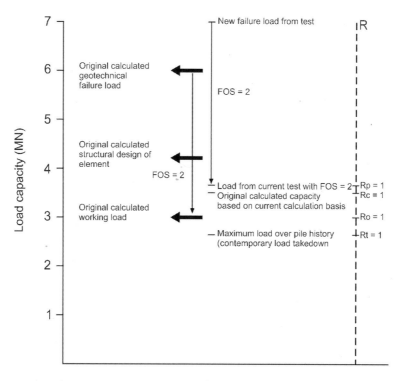

Figure 4.5   *Example illustrating relation of different "R" factors*

If the load take-down value is higher than that calculated using current practice then there could be cause for concern as the original building was loaded higher than new calculations would allow. This is normally related to the foundation operating at a lower factor of safety than would be desirable, but still performing adequately. This was not unusual for old buildings from the 19th century or earlier, constructed with flexible materials, whose foundations had been designed empirically and the dead load added over a period of time. In this situation, foundation reuse would be possible if it is believed that the old structure had worked well and excessive settlement did not occur. An assessment of possible advantages and disadvantages of the different reuse load factors, R, is given in Table 4.4.

Table 4.4   *Features of different calculation approaches for reuse load factor R*

Calculation approach	Reasons why approach might be valid	Reasons why approach might not be valid
Original calculations (R_o)	Every effort would have been made to achieve accurate load estimate at that time.	Available calculations might not have been final and may have been superseded. Current theory may show it had an inappropriate design basis. Design loads may have been too onerous, eg very high live loads and might not have been tested in practice.
Current calculations (R_c)	Based on current understanding of soil and foundation behaviour.	Crucially dependent on constructed dimensions and condition of foundations.
Load takedown (R_t)	Based on loads proven over many years of successful operation. Generally found to be most useful in practice.	Must be based on proven loads, especially live loads. Unless something is know about the history of the building these can be difficult to ascertain.
Foundation testing (R_p)	Based on proven load test.	Foundation tested might not be representative of the entire foundation system, either in dimensions or in previous loading history. In practice load testing is likely to be limited to piled foundations due to the practicality of testing large shallow foundation elements.

The design value of the reuse factor will depend on the amount of caution that is prudent to apply, for the particular use and characteristics of the structure. Caution can be derived from the:

- robustness of the building
- quality of the records available for the existing piles
- amount of investigation to be carried out.

In circumstances where a foundation underperforms and hence experiences large settlements under loading, a robust superstructure will be capable of redistributing loading away from the underperforming foundation to a stiffer, more competent foundation. A robust superstructure constructed on foundations with good as-built records will provide much greater caution for reloading than a less robust superstructure constructed on foundations with poorer quality or incomplete records. In the latter case more intrusive investigations may be required to provide appropriate levels of caution for reloading.

Figure 4.6 shows the appropriate levels of caution for increasing levels of reloading. Greater levels of caution are needed for higher applied loads on reused foundations. Where sufficient caution cannot be derived from the robustness of the structure, the available old pile records or the amount of investigation which is feasible to undertake, additional piles will be required to supplement the existing foundations.

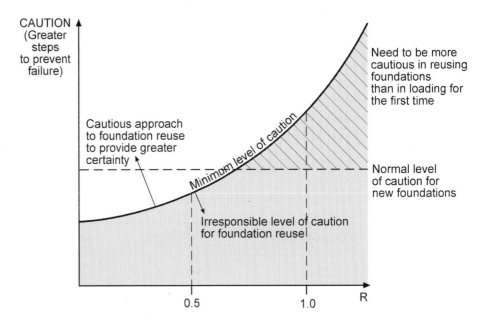

Figure 4.6   *Reuse caution*

In practice, the desired reuse load factor will be dependent upon the loading from the new superstructure and there will be a finite amount of information known about the old piles. An appropriate level of caution will need to be derived from designing the superstructure to be more robust, increasing the extent of the information obtained from the field investigation of the old foundations or the addition of supplemental new foundations. In most cases, the building form will be reasonably fixed, so additional investigations of the old foundations or supplemental new foundations will be the only way of substituting for a lack of information on the old foundations and providing the appropriate confidence.

5   Allowing new foundations to be reused in the future

The best way to avoid stagnating urban sites in the future is to install foundations that can easily be used for successive structures. The reuse potential can be improved through:

- compiling and storing comprehensive records for the construction and operation of the building

- use of instrumentation to record the structural loading and performance of the piles over their lifespan – smart foundations

- use of a foundation platform designed to accommodate a variety of future superstructures.

5.1   Information to be saved

On most sites, saving foundation information is the most valuable approach to enabling future reuse of foundations. The range of information which is valuable includes the parameters identified in Table 5.1.

Table 5.1   *Valuable information to store for new piles*

Site investigation	Design and specification	Construction	Building operation
Factual ground investigation report	Design philosophy and codes	As-built pile location plan	Structural alterations
Geotechnical interpretative report	Geotechnical design	As-built pile schedule, giving diameters and other pile details	Maintenance records
Test results	Structural design of pile elements	Pile construction record sheets	Observations of damage
Records of water level monitoring	Design loads and load combinations applied to each pile	Pile reinforcement schedule	
Archaeological reports	Construction specification	Details of pile integrity testing	
		Results of pile load testing	
		Details of all non-conformances and how they were resolved	
		Confirmation from an independent body that the records are correct	
		Results of settlement monitoring	

There are currently no requirements or guidance on the preservation of records of foundations to aid future reuse. The Construction (Design and Management) Regulations (CDM) 1994 require a Health and Safety file to be compiled to provide information needed during future construction work which includes cleaning, maintenance, alterations, refurbishments and demolition. This file should include as-built drawings for the structure.

Eurocode 7 provides a more detailed framework of records to be kept with regard to the ground condition, design and as-built construction. Eurocode 7 clauses indicated with "P" are in Principle (ie must be followed). For piles, Eurocode 7 specifies that the "installation of all piles is monitored and records are made as the piles are installed" (Clause 7.9 (3)P). "Records should be kept for at least a period of five years after completion of the works. As-built records should be compiled after completion of the piling and kept with the construction documents" (Clause 7.9(5)).

Records relating to pile construction to be saved are to include (Clause 7.9(4)):

- pile number
- pile cross-section and length
- concrete mix, volume of concrete used and method of placing for cast-in-situ piles
- for bored pile, the strata encountered in the borings and the condition of the base if the performance of the pile toe is critical
- obstructions encountered during piling
- deviations of position and direction and as-built elevations.

Eurocode 7 also requires a Geotechnical Design Report to be prepared (Clause 2.8(1)P) including (2.8(3)):

- ground conditions
- proposed construction
- design values of soil and rock properties, including justification, as appropriate
- geotechnical design calculations and drawings
- foundation design recommendations
- note of items to be checked during construction or requiring maintenance or monitoring.

Inspection records during construction are required (Clause 4.2.2(5)P & (6)) including:

- significant ground and ground-water features
- sequence of works
- quality of materials
- deviations from design
- as-built drawings
- results of the measurements and of their interpretation
- observations of the environmental conditions
- unforeseen events
- temporary works including interruptions and their condition on re-commencement.

More important documents should be stored for the lifetime of the relevant structure (Clause 4.2.2(8)).

Typically, design information has been kept at the discretion of the firm of consulting engineers. Likewise, construction records are usually only saved for a finite time by contractors. Space is at a premium and old records are often sorted through and disposed of once the decision has been taken that they are no longer required. Some companies link the preservation of records to their own period of liability – thus records can be disposed of after either six or 12 years, depending on the requirements of their contract. Local building control offices also routinely destroy information to save space.

During pile construction, records are submitted by the main contractor and the design engineer. This usually incorporates the as-built information on a standard form to ensure that construction is complying with design. If kept, this information is a valuable resource for the new design engineer when considering foundation reuse.

As-built information is a much more reliable indicator for the dimensions of the old foundations that remain in the ground than pre-construction design calculations and drawings. Using the as-built information, any deviation from design or defects in the piles can be evaluated for suitability in the new scheme. An extreme situation could occur if piles had been "designed out" during pre-construction discussions with the result that an original design drawing shows a pile in a particular location, but it was never constructed. This could have serious repercussions if pile reuse has been selected as a suitable option and that particular pile was being relied upon as being present beneath a revised pile cap.

To maximise the reuse potential and the resale value of a site, property owners should ensure that good records of the construction and maintenance of their buildings are collected and safely preserved, so that future developments on those sites can benefit from the information. The records should be in a format that can be accessed in the future, so care should be taken when considering electronic formats as both software and hardware are likely to have changed considerably before the records are required. Once compiled, these records should be stored by the property owners in a secure location so that they will not be lost due to future events such as the designer or contractor going out of business or reducing parts of their archives. Storage together with the CDM Health and Safety file is currently recommended as the most appropriate location.

5.2 Smart foundations

Section 4.3 presented the benefits of monitoring an old building, principally to confirm that any obvious structural damage was not due to inadequate foundation capacity during its first design life, and to measure the performance of the old foundations as their loading was removed.

Smart foundations, created through the use of instrumentation, can record the changes in loading and corresponding settlement performance over the lifespan of the foundations. When considering future foundation reuse, this monitoring information will be able to provide additional confidence as to the condition and performance of the foundations and hence is likely to increase the level of reloading that can be assigned for a second design lifespan.

Instrumentation and monitoring systems can be included within the construction of new foundation or added to existing piles after demolition of the previous building and during construction of the subsequent one. These systems may have the function of:

- verifying the settlement performance of the foundations as they are loaded

- confirming that the loads applied to the foundations and the load distribution within the foundations system are within calculated limits

- providing on-going measurements of the loading and settlement of foundations over the service life of the building.

In developing the types of instrumentation to be embedded in piles, prime considerations will be for the instrumentation to be robust with a long operational life to allow the sensors to be interrogated successfully 20–30 years after installation.

The reliability of various instrumentation types has been investigated as part of the RuFUS project along with the instrumentation of new foundations as a real step towards smart foundation. It is hope that some of the aims of smart foundations can be achieved in a practical manner in the near future.

Many building occupiers will perceive a risk to their occupation of a building if monitoring is required post-construction to show that the building is performing as it was designed. For this reason, post-occupation monitoring is tended to be looked upon with suspicion and may not be welcomed. Many developers are reluctant to spend money on monitoring if there is no direct need for it in their building. The benefits of monitoring to improve the reuse potential of the foundations, and hence the resale value of the site, should be emphasised by the design team.

Further developments of the abilities of embedded instrumentation are certain, and will greatly enhance the efficacy of monitoring structural behaviour at less cost. Examples of possible data that may be obtained through instrumentation to create smart foundations in the future are shown in Figure 5.1.

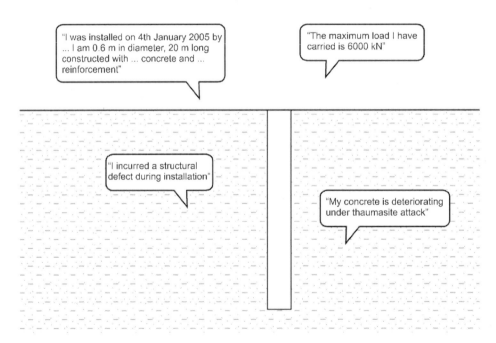

Figure 5.1   *Smart foundations of the future*

5.3 Design of a foundation platform to support successive buildings

A development of the general aim to collect sufficient records for future reuse of foundations currently being installed would be to explicitly design foundations to support a range of different building types, as illustrated in Figure 5.2. This substructure would be designed for a longer design life and to receive a variety of load combinations for a succession of superstructures. While it may be feasible to construct a foundation system that can support a variety of future superstructures, it may not be possible to anticipate future building requirements. Additionally, it is unlikely to be economical to design for all imaginable future requirements. Where there are planning restrictions, such as the maximum building height, the likely maximum structural loading may be more feasible to assess and suitable foundations designed.

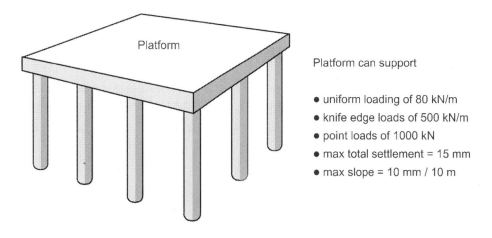

Platform can support

- uniform loading of 80 kN/m
- knife edge loads of 500 kN/m
- point loads of 1000 kN
- max total settlement = 15 mm
- max slope = 10 mm / 10 m

Figure 5.2   *Example of foundation system to support successive superstructures*

6 Questions a developer should ask his team when contemplating foundation reuse

This section comprises a list of types of questions that a developer may wish to ask their professional advisers on projects where reuse of old foundations has been suggested. It is hoped these questions will open up a dialogue about the particular risks inherent in foundation reuse and how they might be mitigated, leading to further questions appropriate for the particular scheme. The text in italics beneath each question gives some background information as to why the question is important.

6.1 For reuse of old foundations

1 What is the range of foundation options and what are the relative risks and costs for each?

   Any response should not downplay the risks, costs and possible delays that could occur if foundations are reused. Extensive investigations may be needed to make sure that the old foundations can safely be relied upon.

2 If a foundation reuse strategy is followed, will the new building sit on foundations that are comparably as reliable as new ones?

   Both new and reused foundations need to be reliable. If the level of uncertainty about performance is higher than normally acceptable, then the suitability of reuse is doubtful.

3 How will the risks be tracked and managed during the foundation reuse design process?

   With foundation reuse, initially there will be many more uncertainties than with conventional foundation design. During the desk study, design and investigation process, these uncertainties should be progressively reduced, so that by the start of construction, all risks should have been successfully mitigated.

4 What is your company's previous experience of reusing foundations and how closely did the final solution match the first proposals?

   In foundation reuse, a high degree of reliance is being placed on professional judgement in a complex area of soil-structure interaction, needing a good mix of structural and geotechnical design skills. It is useful to verify the skills of the individuals on whom such reliance is placed. An approximate measure of the skills and experience is whether a radical change in strategy was needed during the decision-making process on a previous project, ideally any significant changes should have been raised in advance unless they were due to unforeseeable circumstances.

5 If a foundation reuse strategy is progressed, what factors will control whether it will eventually be concluded that reuse is the correct course?

   Normally in foundation reuse, a firm conclusion cannot be reached until some investigation of the old foundations has been completed, which cannot take place until the first building has been demolished. This may affect project cost and programme certainty and may not be resolved until relatively late in the design development programme. It is important for the whole team to have defined for them the factors that may be critical, and the level of certainty which will be required to allow reuse to safely proceed.

6   If it is discovered at the end of the foundation investigation that the risks of reusing some foundations are too high, what would be the new foundation strategy, and what are the likely effects of that new strategy on project cost and programme? What likelihood should be assigned to this scenario?

From the start of the project all foundation options need to be assessed and a fall-back strategy should be maintained. The strategies should include mitigation measures and alternative foundation options to be adopted depending on which risks of reusing some foundations are identified as being too high. It is important that the project contingency should allow for these risks to cost and programme.

7   In the strategy being adopted, which organisation carries the risk of:

- the reuse concept and design not working – either it is not feasible, or the design does not sufficiently mitigate the risks

- the reuse execution not working – ie the concept and design are valid, but the investigations or construction were improperly carried out?

What checks will be enacted to ensure that the strategy will work?

In risk management, it is important to be clear from the outset about responsibilities and ownership of issues. If responsibilities are clear, then appropriate insurances can be put in place. Some checking to verify successful implementation is prudent.

8   Should the completed structure exhibit signs of damage due to a reuse strategy not having worked as intended, how is the damage likely to manifest itself and what repairs might be appropriate?

Foundation reuse involves some risks, and in most cases these can be resolved to give equivalent confidence to new foundations. It is important that the developer should carefully consider the consequences of construction problems for him and his business continuity. For buildings fulfilling critical business functions or where the consequence of a problem would carry prohibitive financial implications, then a more robust approach should be adopted against risk of structural failure. This applies as much to other aspects of the construction as it does to foundation reuse.

The risk should be low, but there may be small relative settlements, as there would be with a new-build scheme. There may be circumstances where the consequences of larger relative settlements are minor and the developer is prepared to accept this occurring.

6.2   For new foundations to be reused in the future

9   What steps are being taken to ensure that any new foundations can be easily reused to support a new building in the future?

Building owners are being encouraged to keep records that will allow new foundations to be reused in the future. Such records may prove invaluable the next time the site is redeveloped, allowing a more efficient building to be built more quickly. Initiatives such as this guide and the RuFUS project are highlighting the benefits of the option to reuse. Research on how foundations act during demolition and subsequent re-loading is being undertaken at several sites and will allow design engineers to better predict the capacity of reused piles.

10 How will comprehensive records on these new foundations be safely preserved so that they will be available for the fresh building?

At present no system exists for the preservation of foundation records. Records can be retained by the owner, firm of consulting engineers or even the local building control office. However, wherever they are saved it is important that the owner is aware of them and that they can be accessed as required. It would also be beneficial if, during the life of the building, any structural alterations are made to the building that a record of this is added to the original information. The finding of the RuFUS project includes recommendations for a documentation system.

7   References

7.1   References cited from text

Ackroyd, P (2001)
London, a biography
Vintage (ISBN: 978-0-09942-258-7)

BCA/BSRIA (1999)
Refurbishment of concrete buildings: structural and services options
Guidance Note GN8/99, The Building Services Research and Information Association,
Bracknell, UK

Bailey, N (1981)
Fitzrovia
Historical Publications Ltd, Herts (ISBN: 978-0-95036-562-6)

Biddle, M (1994)
"What future for British Archaeology"
In: *Proc opening address 8th Annual Conf. Institute of Field Archaeologists, Archaeology in
Britain Conference*, Bradford 13–15 April 1994, Oxbow Books, Oxford

Building Research Establishment (2005)
Concrete in Aggressive Ground
BRE Special Digest 1. Garston: Building Research Establishment (superseded BRE
Digest 363 which superseded BRE Digest 250) (ISBN: 978-1-86081-754-0)

Butcher, A P, Powell, J J M and Skinner, H D (eds) (2006)
Reuse of foundations for urban sites – A best practice handbook
BRE Press (ISBN: 1-86081-938-9)

Butcher, A P, Powell, J J M and Skinner, H D (2006)
"Reuse of foundations for urban sites"
In: *Proc Int. Conf on Reuse of Foundations for Urban Sites*, 19-20 October 2006
BRE Press (ISBN: 1-86081-939-7)

Chapman, T J P, Marsh, B and Foster, A (2001)
"Foundations for the future"
In: *Proc ICE, Civil Engineering*, **144**, February 2001, pp 36–41

Chapman, T J P, Chow, F C and Skinner, H (2002)
Building on old foundations – sustainable construction for urban regeneration
CEWorld Conference, ASCE on-line conference <http://www.ceworld.org/>

Chapman, T J P and Marcetteau, A. (2004)
"Achieving economy and reliability in piled foundation design for a building project"
The Structural Engineer, June 2004

Chow, F C, Chapman, T J P and St John, H D (2002)
"Reuse of existing foundations: planning for the future"
In: *Proc 2nd Int. Conf. Soil structure interaction in urban civil engineering*, Zurich

Chow, F C, Jardine R J, Nauroy J F and Brucy F (1997)
Time-related increases in the shaft capacity of driven piles in sand
Géotechnique, 1997, **47**, No.2, pp 353–361

Davis, M, Hall, A, Kenward, H and Oxley, J (2002)
"Preservation of Urban Archaeological Deposits: monitoring and characterisation of
archaeological deposits at Marks and Spencer, 44-45 Parliament Street, York"
Internet Archaeology, <http://intarch.ac.uk/journal/issue11/oxley_index.html> (ISSN:
1363-5387)

Department of the Environment (1990)
PPG16 – Planning Policy Guidance: Archaeology and planning
London: HMSO

Nixon, T (1998)
"Practically preserved: observations on the impact of construction on urban
archaeological deposits"
In: *Proc conf M Corfield, P Hinton, T Nixon and M Pollard (eds) Preserving archaeological
remains in situ* 1–3 April 1996. Museum of London Archaeology Service, London, pp
39–46

St.John, H D, Chow, F C and Harwood, A (2000)
"Follow these footprints"
Ground Engineering, Dec, pp 24–25

Wardle, I F, Price, G and Freeman, T (1992)
"Effect of time and maintained load on the ultimate capacity of piles in stiff clay"
In: Sands (ed) *Piling: European practice and worldwide trends , 1992, pp 92–99*
<http://www.arup.com/_assets/_download/download164.pdf>

Whitaker, T and Cooke R W (1966)
An investigation of the shaft and base resistance of large bored piles in London Clay
In: Proc Symposium *Large Bored Piles, 1966, ICE, London, pp 7–49*

7.2   Codes, Standards and Regulations

BS 8004:1986 *Code of practice for Foundations* – (formerly CP 2004)

NA to BS EN1990:2002 *Basis of structural design*, UK National Annex for Eurocode 0,
London

BS EN1997-1:2004 *Eurocode 7: Geotechnical Design – Part 1: General Rules*

BS EN1990:2002 *Basis of structural design*, London

New York City Department of Buildings *Building Code of the City of New York*, (2003)

The Construction (Design and Management) regulations 1994 (as amended), HMSO (1994)

7.3   Other useful foundation reuse references

Anderson, S, Bird, J and Chapman, T (2006)
Assessment of risks and opportunities of foundation reuse
In: *Proc. Conf Re-use of foundations on urban sites,* London

Chapman, T J P, Butcher, A and Fernie, R (2002)
A generalised strategy for reuse of old foundations
In: *Vanicek, I, Barvinek, R, Bohac, J, Jettmar, J, Jirasko, D, and Salak, J (eds), Proc. XIII European Conference Soil mechanics and geotechnical engineering,* **3**, Czech Geotechnical Society, Prague, 2003

Paul, T, Chow, F C and Kjedstad, O (2002)
Hidden aspects of urban planning, surface and underground development
Thomas Telford, London

Mitchell, J M, Courtney, M and Grose, W J (1999)
Timber Piles at Tobacco Dock, London
In: *Proc ISSMFE conf Colloque International Fondations Profondes,* Paris, March 1999

Coles, B, Henley, R and Hughes, R (2001)
The reuse of pile locations at Governor's House Development Site, City of London
In: *Proc. Int conf Preserving archaeological remains in site,* Museum of London Archaeological Service and University of Bradford, September

Thornburn, S and Thornburn, J Q (1977)
Review of problems associated with the construction of cast-in-place concrete piles
CIRIA Report PG2, London (ISBN: 978-0-86017-020-4)

A1   The RuFUS project

The Reuse of Foundations for Urban Sites (RuFUS) is a European Union partially-funded research project to set guidelines that will allow foundations to be reused more often. The project consists of eight European organisations working in partnership:

Country	Organisation
UK	BRE (project co-ordinator)
	Arup
	Cementation Foundations Skanska
France	Soletanche-Bachy
Germany	Technical University of Darmstadt
	Federal Institute for Material Testing and Research (BAM)
Greece	Stamatapoulos Associates
Sweden	Swedish Geotechnical Institute

The project started in early 2003 and culminated in an international conference held on 19–20 October 2006 at BRE, Watford, UK.

Through the publication of a best practice handbook at the 2006 conference, the RuFUS project has provided:

- a decision model for end-users and their professional advisors that will include economic and environmental impact and risk assessments for the reuse of foundations. This will demonstrate to clients, owners, insurers and regulators the economic and environmental advantages of the reuse of foundations, as well as the control of safety and minimisation of liability

- a documentation system/database to record foundation design and construction details, in a standard format, to be readily available to subsequent site owners/construction professional for the reuse of foundations during redevelopment of the site

- improved non-destructive testing techniques to assess the dimensions and reliability of the old foundations under consideration for reuse

- the basis for a foundation behaviour system that will use the unloading during the demolition of the initial superstructure to predict the foundation behaviour during reloading for a different load distribution. This will benefit designers considering foundation reuse by proving the performance of the foundation they are using for the replacement building

- reliable instrumentation for "smart" foundations to monitor foundation behaviour during use for direct comparison with the design behaviour. This will show the in-service performance of the foundation, the distribution of loads and the change in distribution with time. This data (in conjunction with better preservation of piling records) will serve to facilitate any future reuse during redevelopment of the site.

The accompanying conference will demonstrate a range of projects where foundation reuse has been considered a part of a redevelopment foundation strategy.

The project website can be accessed on <http://www.webforum.com/rufus/home/>.

A2   Case studies

The following is a selection of case studies of completed and ongoing projects involving foundation reuse.

Case study A2.1   **Thames Court** *(courtesy Geotechnical Consulting Group)*

New piles were stitched into the existing raft to carry local increases in loading as the structural performance of the raft may have been insufficient. Use of the raft avoided problems that would have been caused by excavation with a high water table.

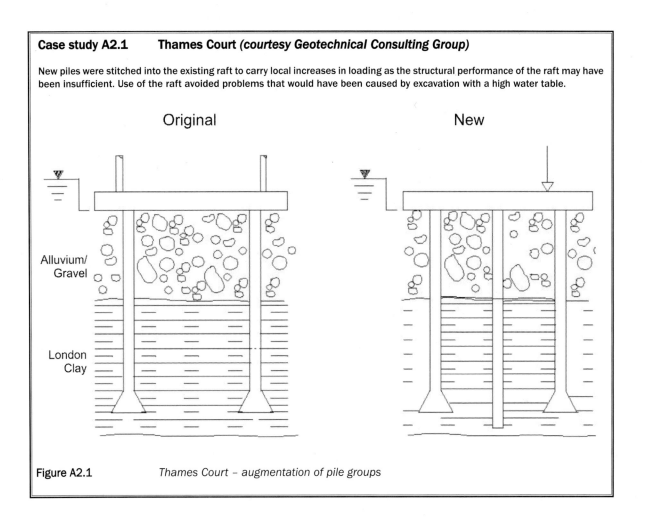

Figure A2.1   *Thames Court – augmentation of pile groups*

Case study A2.2   Empress State Building, Earls Court, London *(courtesy Geotechnical Consulting Group)*

This project involved the refurbishment of an existing 28 storey building, adding extra floors, a revolving restaurant and an extra line of columns to the south façade. It was determined that this would result in an addition of around 30 per cent of load to the existing under-ream piles beneath the south façade.

Original designs and an original layout plan were available. The solution adopted was to create a stiff beam linking the old piles to a new foundation system which would also carry the new line of columns. This was intended to reduce the risk of unacceptable settlements occurring to the old piles.

Figure A2.2
*Empress State Building
– original structure*

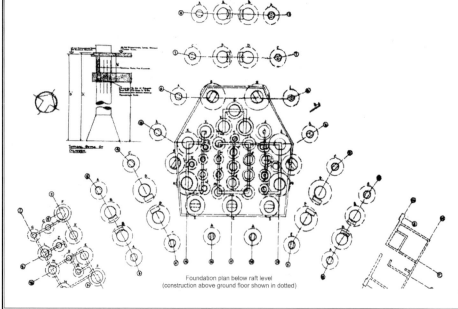

Foundation plan below raft level
(construction above ground floor shown in dotted)

Figure A2.3
*Empress State Building
– original piling layout
(from record drawings)*

Case study A2.2   Empress State Building, Earls Court, London *(courtesy Geotechnical Consulting Group) (contd)*

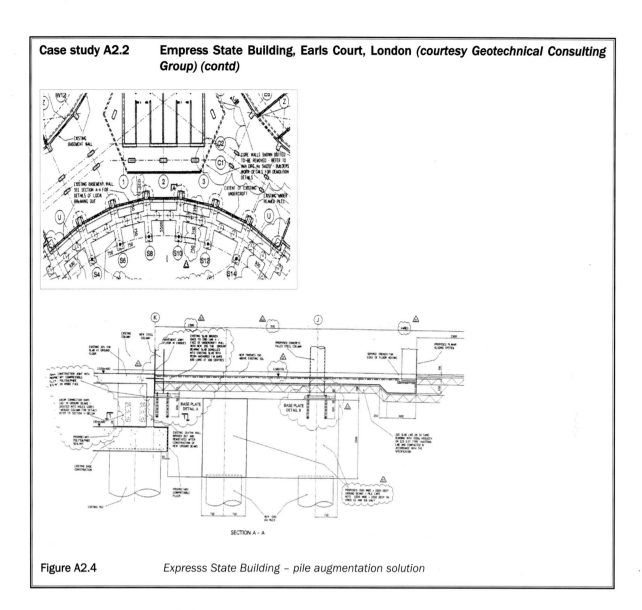

Figure A2.4   *Expresss State Building – pile augmentation solution*

Case study A2.3   Juxon House, St Paul's Churchyard, London *(courtesy whitbybird and Geotechnical Consulting Group)*

This project involved interaction between old, mainly under-reamed piles with base diameters of up to 3.2 m, and new piles.

The original design loads were confirmed through whitbybird's load take-down for the existing building. The original design had been substantially modified on site, as a scour hole in the London Clay had been found, but fortunately good records of the changes made on site were available.

New piles were required where there were no existing piles as fill over relatively thin Terrace Gravel made raft solutions problematic. The locations of existing piles and the presence of archaeological remains constrained the new pile locations. In some areas, piles were installed to supplement existing piles where loads were higher than those previously carried.

The new piles were 900 mm diameter CFA piles, designed both to carry the anticipated extra loads but also to provide compatible settlement performance.

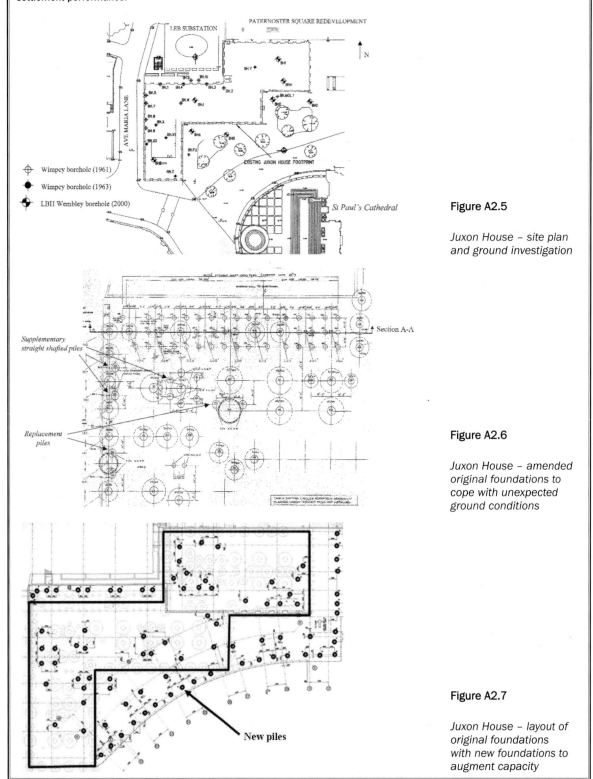

Figure A2.5

Juxon House – site plan and ground investigation

Figure A2.6

Juxon House – amended original foundations to cope with unexpected ground conditions

Figure A2.7

Juxon House – layout of original foundations with new foundations to augment capacity

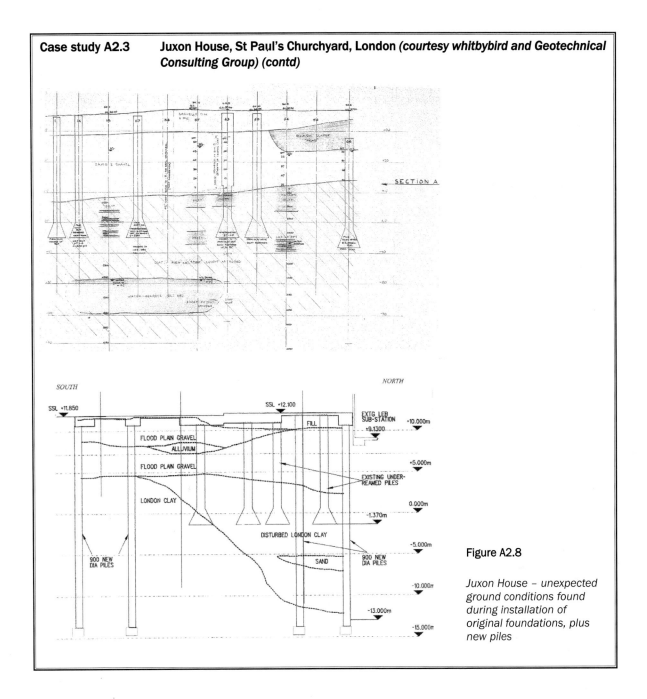

Figure A2.8

Juxon House – unexpected ground conditions found during installation of original foundations, plus new piles

Case study A2.4 Reconstruction of Network Rail Bridge 2/11 ~ 2/12, adjacent to Clapham Junction (*courtesy Mott MacDonald*)

Bridge 2/11 ~ 2/12 is a skewed bridge structure consisting of two spans each carrying 13 decks which support thirteen railway lines carring mainly passenger services to and from the south of England into London Waterloo and London Victoria stations. Below, the structure spans over four railway lines which carry predominately freight traffic. The site is situated approximately one mile north-east of Clapham Junction station.

Bridge 2/11 ~ 2/12 was originally built circa 1890 and is comprised of mass brick abutments and a central pier that supported a deck of wrought iron plate girders.

The existing bridge decks were in poor condition with ongoing maintenance liabilities and it was decided to replace the superstructure. The new superstructure would lead to changes in loading on the substructure and foundations, despite the weight of the new bridge decks being kept to a minimum. An appraisal was carried out on the feasibility of foundation reuse, comprising:

1 Desk study, including search of old foundation records.

2 Walkover survey of exposed substructure/superstructure.

3 Detailed load analysis of the existing bridge configuration.

4 Coring through the substructure/foundations to measure key dimensions and to appraise the extent of material deterioration.

5 Trial pits to evaluate the near surface ground conditions.

The bridge abutment and piers are masonry founded on strip footings. Although the studies provided some ambiguous data, it was possible to make conservative estimates of key dimensions for the assessment of existing foundation capacity based on a knowledge of Victorian railway construction practice. The comparison of load take-downs between the new and old structures revealed that:

● it would be possible to avoid an increase in stress on the existing bridge abutment foundations, if the abutment bearings were relocated to provide a more efficient redistribution of load

● the new structure would lead to an increase in bearing pressure of about 30 per cent at the central bridge support pier.

The increase in bearing pressure at the bridge pier necessitated a more detailed evaluation of the load bearing characteristics of its foundations, together with more detailed investigation of the foundation dimensions and adjacent ground conditions. This detailed evaluation indicated that the settlement under the increased foundation load would be acceptable.

The key drivers for the project were:

● the rail locked nature of the site

● the limited space and severe access constraints

● all reconstruction works had to be completed within a limited track possession.

These drivers meant that there were significant advantages in the reuse of existing foundations. The inspections of the substructure revealed that there were some defects, but that these could be repaired by relatively minor works. There was no evidence of substantial settlement of the existing foundations.

Following successful reconstruction of the bridge, its subsequent performance has been satisfactory. This is one of many applications of foundation reuse for railway bridges, where the reuse of foundations dating from about 100 to 150 years ago is relatively common.

Figure A2.9 *Bridge 2/11 ~ 2/12, demolition of existing decks*

Figure A2.10 *Bridge 2/11 ~ 2/12, bridge reconstruction nearing completion*